REFERENCE GUIDE TO USEFUL ELECTRONIC CIRCUITS AND CIRCUIT DESIGN TECHNIQUES

KERWIN MATHEW

REFERENCE GUIDE TO USEFUL ELECTRONIC CIRCUITS
AND CIRCUIT DESIGN TECHNIQUES

PREFACE

This book has been prepared for practically all types of readers, ranging from laymen to engineers. There are a number of descriptions of interesting electronic circuits in the book which should be useful to the practitioner in the field. There are also some electronic circuit design techniques brought up for the benefit of the reader.

The book with its special selection of electronic circuits would serve as a reference guide for the practicing engineer and the student of electronics.

Kerwin Mathew, Ph.D., PE, CMfgT, CPM

CONTENTS

1 AC LINE FILTERS

Harmonic filters can be linked to the neutral from each line to attenuate the penetration of harmonics into the AC system from a rectifier load.

The diagram below shows harmonic filters that enable the harmonic currents to be bypassed:-

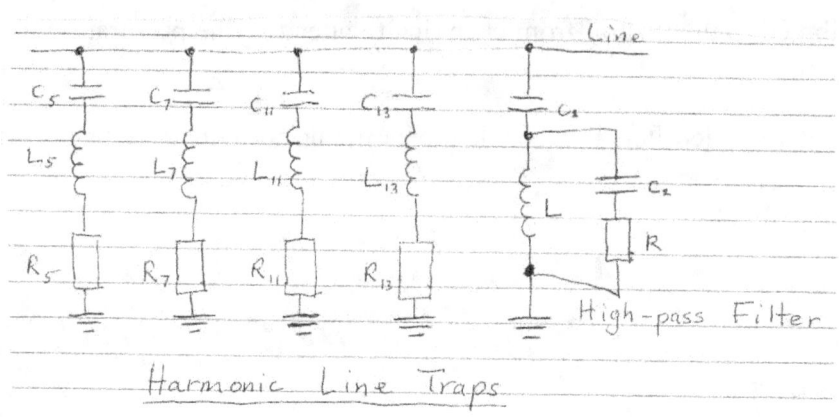

Harmonic Line Traps

For the fifth, seventh, eleventh and thirteenth harmonic components in a six-pulse system, tuned harmonic filters are provided. A high-pass filter is provided for the higher-order harmonics. To prevent excessive loss at the fundamental frequency, care should be taken. A problem that is often encountered is that of frequency drift, which could be as high as +/- 2 per cent in a public supply system. To be effective the filters have either to have a low Q factor or be automatically tuned.

Normally, very large converters, e.g., those used in HVDC transmissions, will use this kind of filters.

2 AC POWER CONTROL CIRCUITS

A control circuit is an arrangement of physical components that are connected or related in such a way as to command, direct or regulate itself or another system.

In the AC power control circuit below, a sinusoidal waveform is applied at the input:-

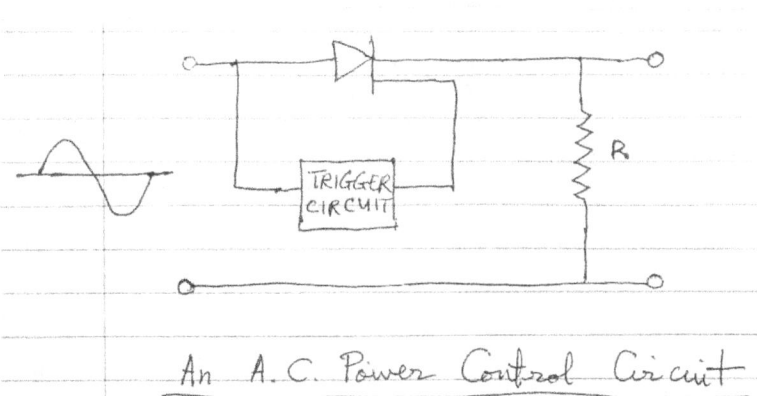

An A.C. Power Control Circuit

The output waveform of this circuit would be as follows:-

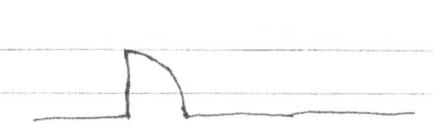

3 ANALOG-TO-DIGITAL CONVERTERS

Though the reading of a modern measuring instrument is frequently in digital form, the input from, e.g., a transducer, is in analogue form. Thus an analogue-to-digital (A/D) converter is needed.

A/D Conversion
Below is the block diagram for a four-bit "counter" type A/D conversion circuit:-

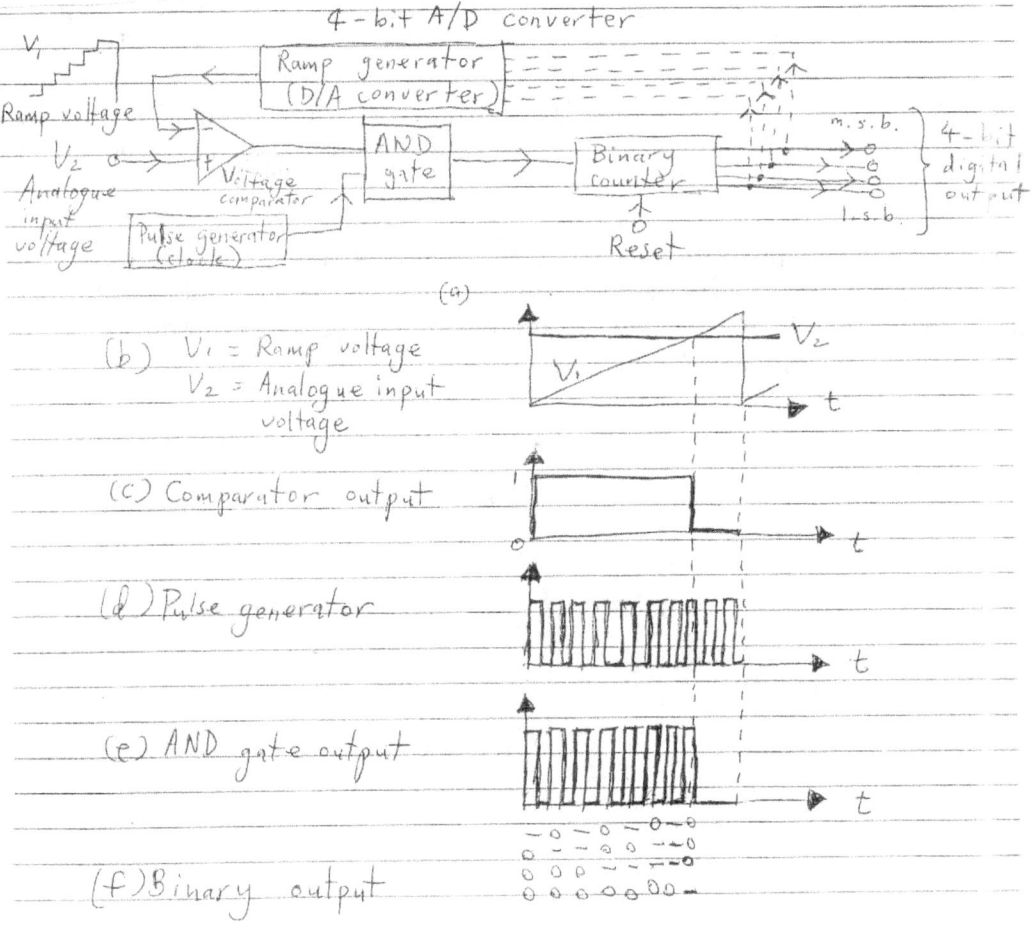

4-bit A/D converter

V_1 Ramp voltage

Ramp generator (D/A converter)

V_2 Analogue input voltage

Voltage comparator

Pulse generator (clock)

AND gate

Binary counter

Reset

m.s.b.

l.s.b.

4-bit digital output

(a)

(b) V_1 = Ramp voltage
V_2 = Analogue input voltage

V_1 V_2 t

(c) Comparator output t

(d) Pulse generator t

(e) AND gate output t

(f) Binary output

In the above analogue-to-digital converter, an operational amplifier (op amp) is used as a voltage comparator. V_2, the analogue input voltage (which is a steady dc voltage), is applied to the non-inverting (+) input. A ramp generator with a repeating saw-tooth waveform voltage V_1 supplies the inverting (-) input.

The comparator's output is applied to one input of an AND gate. It is "high" (one) until V_1 equals (or exceeds) V_2, when it goes "low" (zero). A steady train of pulses from a pulse generator (known as a "clock") supplies the other input of the AND gate. The AND gate opens and gives a "high" output, i.e., a pulse, when both these inputs are "high". The number of pulses obtained from the AND gate depends on the "length" of the comparator's output pulse, i.e., on the time V_1 takes to reach V_2. If the ramp is linear, this time taken is proportional to the analogue voltage. The binary counter records the output pulses from the AND gate; these output pulses are the digital equivalent of the analogue of the analogue input voltage V_2.

The ramp generator is a digital-to-analog (D/A) converter. It gets its digital input from the binary counter (this digital input is represented by the dashed lines shown in the above block diagram). The binary counter advances through a normal binary sequence. As it does so, a staircase waveform with equal steps (i.e., a ramp) is formed at the output of the D/A converter.

4 AUDIO POWER AMPLIFIERS

Amplifiers that supply power to a loud-speaker can be either of the discrete circuit type or the integrated circuit (IC) type. While the latter type has all the advantages of being in integrated form, it faces the problem of getting rid of unwanted heat which makes its design more difficult. This heat dissipation problem can be overcome by mounting a heat sink on the IC according to the manufacturer's recommendations.

The circuit below utilizes a LM380 power amplifier:-

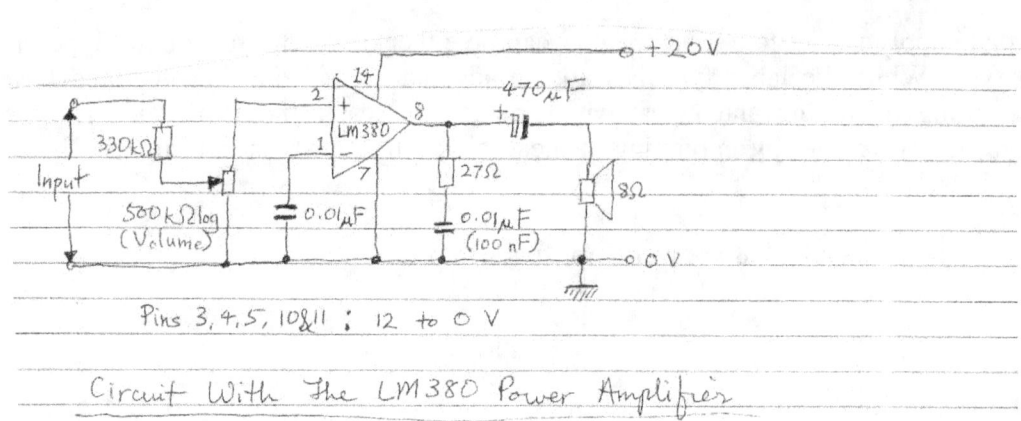

Pins 3, 4, 5, 10 & 11 : 12 to 0 V

Circuit With The LM380 Power Amplifier

The LM380 power amplifier is a 14-pin d.i.l. amplifier with a typical 2W output into an 8 Ω loud-speaker. It has a supply of 20 V and a heat sink (comprising of about six square centimeters of metal strip) connected to the center pins on its d.i.l. package. It also has protection circuits which operate if the IC's temperature becomes too high or it is overloaded (e.g., when the speaker has an impedance that is too small and a large input).

The LM380 has a bandwidth of 100 kHz and a fixed-loop gain of 50. The non-inverting input is connected internally to ground through a 150 k Ω resistor. The input could be taken from the output of a LM381 preamplifier circuit or directly from a ceramic or crystal pick-up or microphone. To produce uniform amplification over a wide range of frequencies a few extra components are required.

5 BAND-ELIMINATION LC FILTERS

The band-elimination filter makes use of the different impedance characteristics of parallel- and series-resonant LC circuits. At the resonant frequency, a parallel LC circuit achieves maximum impedance, while a series LC circuit has minimum impedance. The band-elimination filter circuits below comprise these two LC circuits:-

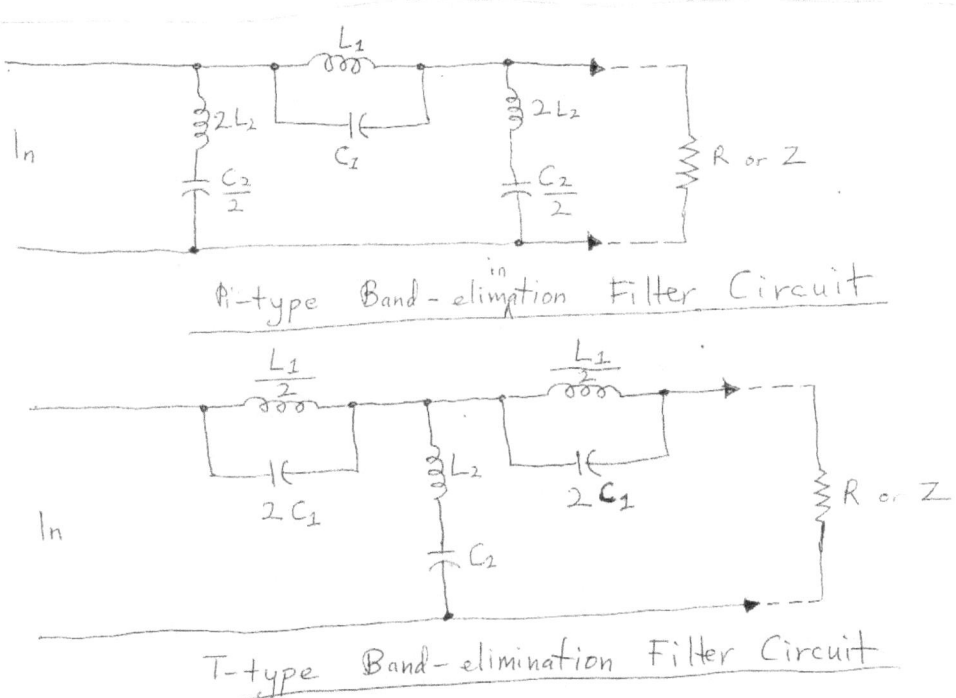

Π-type Band-elimination Filter Circuit

T-type Band-elimination Filter Circuit

At the center frequency of the desired band, there is minimum impedance at the series arm. On either side of the resonance, the impedance increases. This results in the center frequency being bypassed. At the center frequency, there is maximum impedance at the parallel arm, while impedance decreases on either side of the resonance. This eliminates the center frequency as well as a band of frequencies on either side of the resonance.

6 CLAMPING CIRCUIT

When it is necessary to fix the peak value of some recurring waveform to a reference level, the clamping circuit is used. That is, the output signal has a dc component introduced into it. Such a circuit, having a diode, and commonly used, is shown below:-

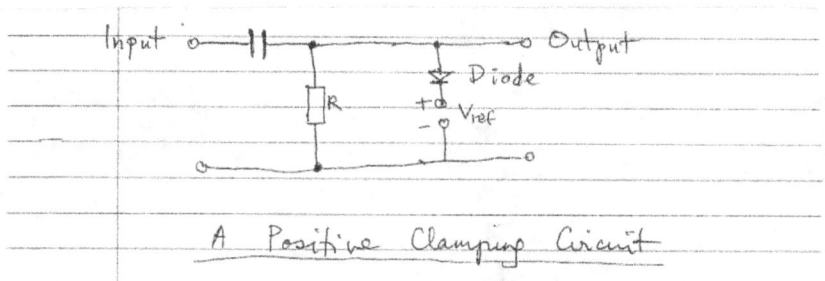

A Positive Clamping Circuit

When the input waveform is at its most positive excursion and effectively clamps the output signal to a dc voltage of $V_R + V_D$, where V_D is approximately 0.6 V for a silicon diode, the diode conducts for a short period. Compared with the periodic time of the lowest signal frequency of the input, the input time constant CR should be long.

By reversing the diode and the polarity of the reference voltage, a clamp to some negative voltage level could be achieved.

The output waveform is clamped to nearly zero volt if the reference voltage is zero (the diode is connected to ground). Such a circuit is commonly called a dc restorer.

7 CLOSED LOOP (NEGATIVE FEEDBACK)

Negative feedback is used by most amplifiers and control systems to stabilize the gain, to reduce non-linearity and distortion, and, to improve frequency response. The amplifier or control system could be said to be in a closed loop condition when a feedback loop is connected to feed a portion of the output back to oppose the input signal. The following is the closed loop voltage gain of an amplifier with negative feedback:-

$A_{VCL} = A_{VOL}/1 + A_{VOL}\beta$

The condition where no feedback signal is applied and the gain of the system is at maximum uncontrolled value is "open loop". Open loop in a control system would give an output which bears no relationship to the effect it produces. A closed loop system is actuated by error and automatically adjusts the output to give a fixed effect.

In a motor speed controller with an open loop, the input sets the power applied to the motor and no adjustment is made for loading effects. The speed of the motor would vary with the load. With a feedback loop to give closed loop control, the power applied to the motor could be altered to suit changing loads. This keeps the motor's speed relatively constant. A transducer, e.g., a tachogenerator, senses the speed of rotation of the motor's shaft. It feeds back a voltage proportional to the speed. This is compared with the input reference and the final output is automatically adjusted to the required level.

8 CRYSTAL OSCILLATORS

Basically, an oscillator is a timing reference. High accuracy is often required of this reference. Long-term frequency stability is a function of the constancy of the elements that make up the oscillator (or, resonator, as it is sometimes called). However, the feasibility of using oscillators is restricted to low frequencies.

The piezoelectric effect is nature's mechanism for converting mechanical energy into electrical energy and vice versa. The crystalline materials of crystal oscillators have mechanical pressure which causes electric charges to be displaced. This results in the production of an electrical voltage. On the other hand, a voltage brings about mechanical pressure in the crystal oscillator. A crystal oscillator could be made in such a manner that a desired resonant frequency is obtained. Such an oscillator could have a very high Q-factor. The frequency range of the quartz crystals in such oscillators is normally 20,000 hertz to 200,000 hertz.

The construction of a crystal oscillator is shown below:-

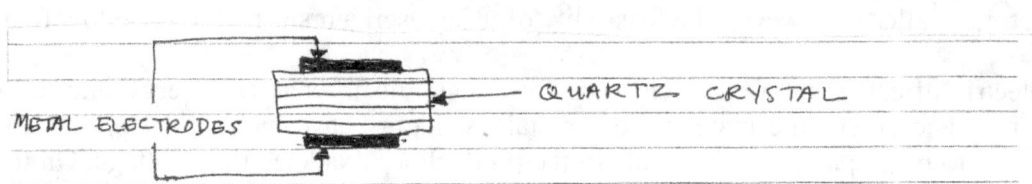

A mechanical shear stress appears in the quartz crystal if a voltage is applied to the metal electrodes and vice versa.

Below is the equivalent circuit of a crystal oscillator:-

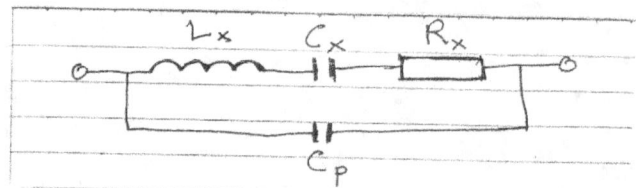

In the above circuit, both series resonance and parallel resonance are evident. Both resonant frequencies are close to one another because $C_X \ll C_P$ always. Both these types of resonance could be found in oscillator and filter circuits. The thickness of the sheet of crystal determines the resonant frequency.

However, the construction of a crystal oscillator is not without problems. If the fundamental frequency were used and for frequencies exceeding about 20 MHz the sheet of crystal would be unpractically thin. Also, because of the dissipation in the crystal and non-linear behavior at large excitations the maximum tolerable voltage is restricted.

9 THE CYCLOCONVERTER

Cycloconversion mostly involves the direct conversion of energy to a different frequency by synthesizing a low-frequency wave from appropriate sections of a high-frequency source.

Circuits which have a dc source are broadly covered by inverters. These inverters enable an alternating voltage to be synthesized for feeding to an ac load by appropriately switching the rectifying devices.

A cycloconverter normally comprises of two converters connected back to back as shown below:-

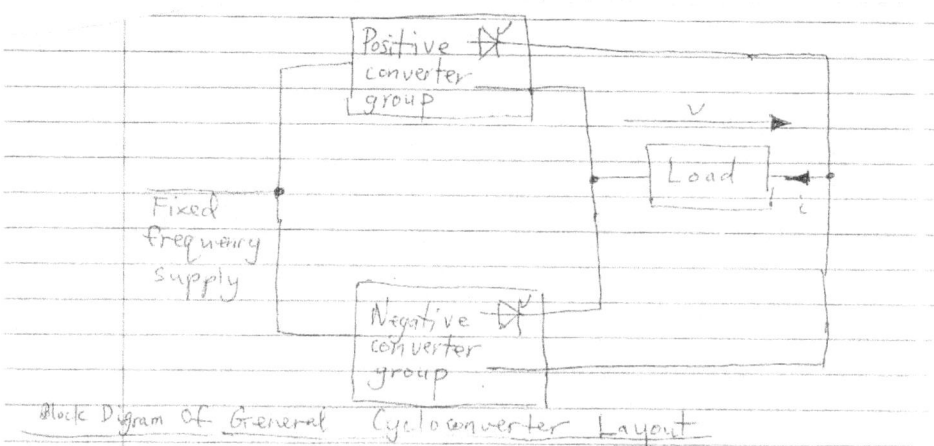

Block Diagram Of General Cycloconverter Layout

Generally, the instantaneous power flows in the load belong to one of four periods. The diagram below shows these load waveforms:-

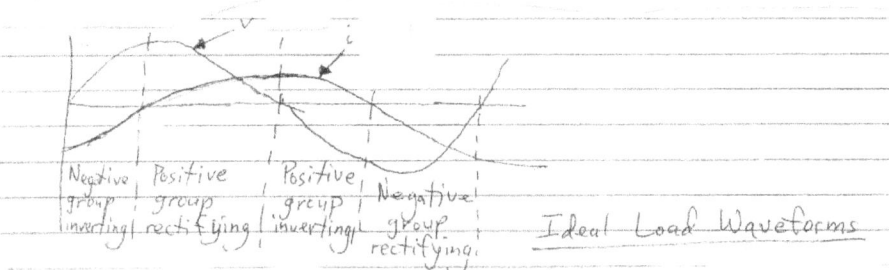

Negative group inverting | Positive group rectifying | Positive group inverting | Negative group rectifying.

Ideal Load Waveforms

Power will flow into the load during the two periods when the product of load voltage and current is positive. This gives rise to a situation where the converter groups rectify the positive and negative groups conducting respectively during the appropriate positive and negative load-current periods.

The times when the product of load voltage and current is negative are represented by the other two periods, whence the power flow is out of the load. The converters will now operate in the inverting mode.

10 DIGITAL NOISE REDUCTION (DNR)

A noise reduction system is needed when there is the presence of a picture field store in a TV receiver or monitor. Weak signal reception and tape noise in VCR playback are some of the many reasons why noise intrudes on an analogue video signal. The noises of a still video picture have total correlation, though noise is generally random in nature and there is no correlation between successive video fields. The noise cancels to virtually zero if several video fields are integrated, while the picture-signal components add.

In a field-store TV set by progressive integration of the present and previous fields, electronic noise reduction could be carried out. Below is a diagram showing how this is done:-

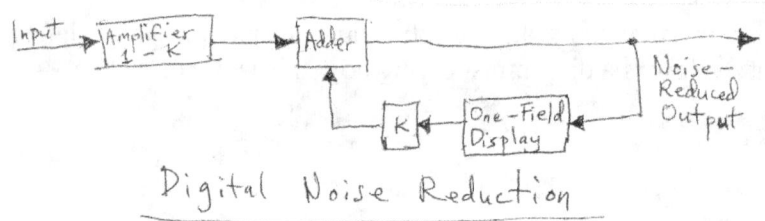

Digital Noise Reduction

In the above diagram, the noise-reduction factor is determined by K, which is a recursive filter. The greater the noise reduction, the higher the K factor. A motion detector which continually adjusts the K factor governs the recursive filter. To prevent blurring due to integration, the K factor is reduced for moving parts of the image. For stationary picture areas, it is increased. The better and more noise-free the picture appears, the longer an object in it remains still.

11 THE FORWARD CONVERTER

The forward converter is a DC-AC-DC converter. The circuit diagrams below show two kinds of forward converter.

Transformers are used by forward converters. The drive to the transformer from the DC input could be a single switch or it could be a push-pull arrangement of two switches. In each case, the circuit does not depend on the storage of energy in the transformer. As is the case in a transformer in a linear power supply, energy is received by the output rectifier at the same time that it is delivered to the primary.

The following points should be noted:-

[a] If the drive is not push-pull, there would be residual magnetism (energy) left on the core. This problem is eliminated by introducing a spare winding and another diode switch into the circuit, which would return the residual energy to the source.
[b] The forward converter circuit arrangement allows isolation of one side from the other. (In many countries, this is required by law in mains supplies.)
[c] The forward converter could develop with equal ease either smaller or larger voltages, compared to the input.
[d] The transformer forces energy to come in "packets"; energy stored in the inductive element is not relied on; and the transformer could thus be responsive to load changes.

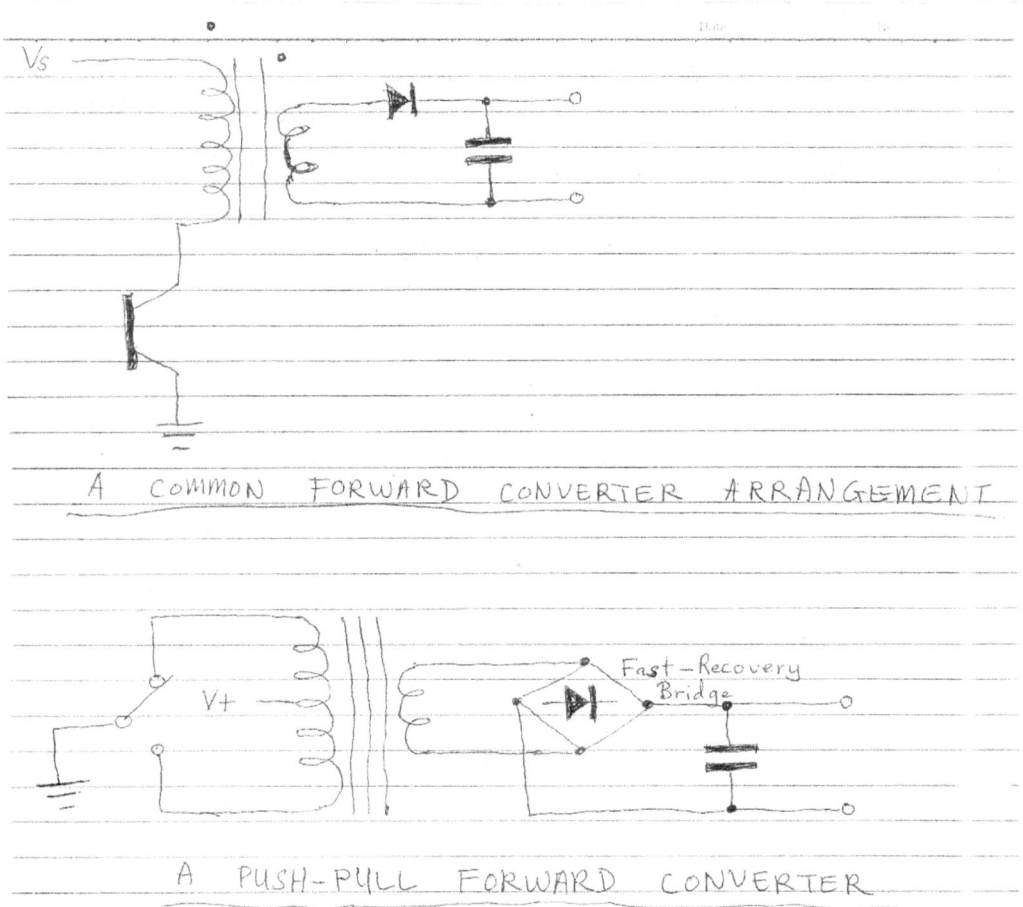

V_S

A COMMON FORWARD CONVERTER ARRANGEMENT

$V+$

Fast-Recovery Bridge

A PUSH-PULL FORWARD CONVERTER

12 FREQUENCY DOUBLER

To resonate at the second harmonic frequency of the input signal, an amplifier with a tuned circuit in the output could be utilized. The amplifier would change the output to twice the frequency of the input.

Using a CMOS exclusive-OR gate type 4070 and an RC network, a digital frequency doubler could be formed. The output from the exclusive-OR would be twice the input frequency when the RC time constant is set approximately equal to one half the input waveform's periodic time, as an exclusive-OR only provides an output when one input is high and no output when both inputs are high.

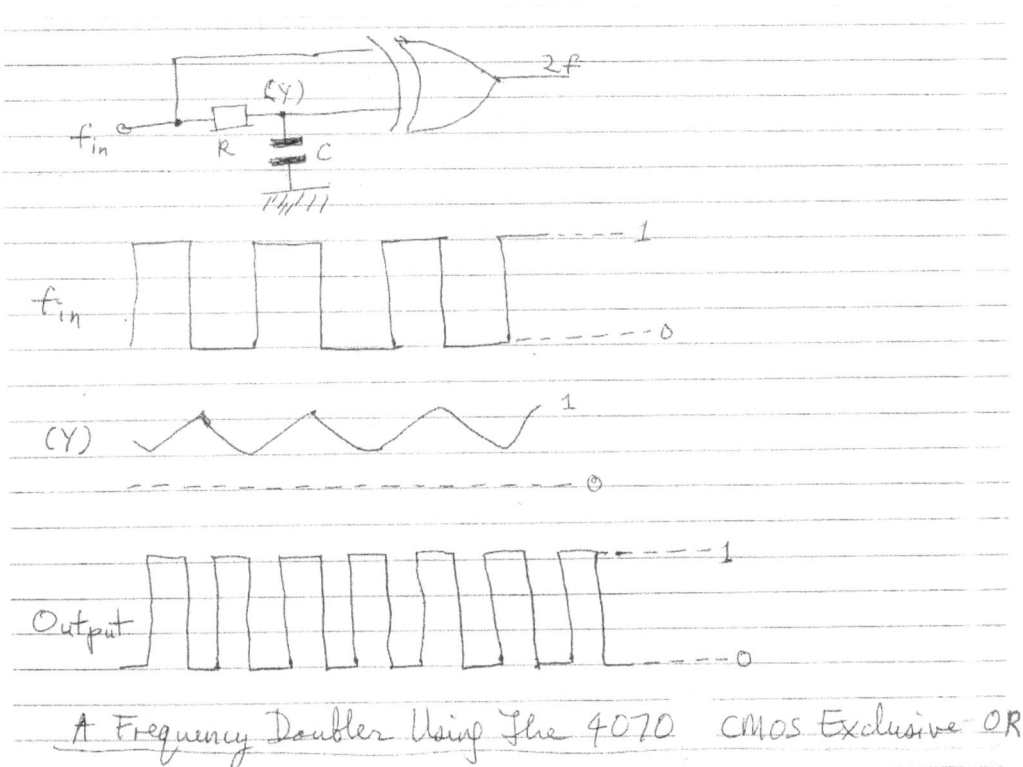

f_{in}

(Y)

Output

A Frequency Doubler Using The 4070 CMOS Exclusive OR

In the above circuit, there is a moment when both inputs are high causing the output to go low, followed by a short time when the input is low, as the signal at (Y) is a delayed version of the input. However, a second output pulse is generated because the level at (Y) is high. Short-duration pulses are generated at the gate output for every edge of the input waveform if the RC time constant is made shorter.

13 THE FREQUENCY DOUBLING CIRCUIT

A frequency doubler can be built to give an output pulse train whose frequency is twice that of a squarewave input signal by using a single 4069 hex inverter IC.

Let us look at the circuit diagram shown below:-

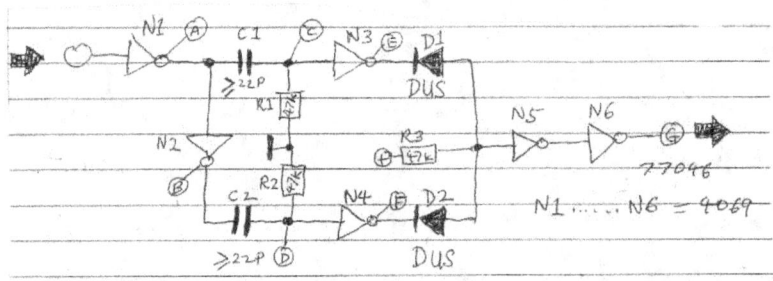

A squarewave signal with a duty-cycle of approximately 50% at a level compatible with CMOS logic (3 - 15 V peak-to-peak depending on supply voltage) is applied to the input at N1. N1 buffers and inverts this input signal, which is inverted again by N2. The outputs of N1 and N2 (shown in A and B below) are squarewave signals that are 180° out-of-phase:-

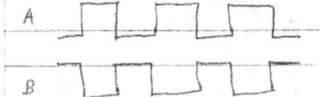

C1 and R1 differentiate the output of N1, while C2 and R2 differentiate the output of N2, resulting in two spike waveforms (shown in C and D below) that are 180° out-of-phase:-

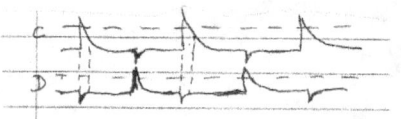

N3 and N4 will then buffer, invert and shape these signals to give the waveforms (E and F) shown below:-

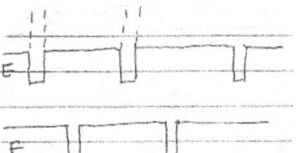

A NOR gate consisting of D1, D2, R3 and N5 then combines these signals, which are finally inverted by N6 to give the output waveform shown in G below:-

The frequency doubling circuit operates over a wide frequency range. The width of the negative-going pulses E and F has to be greater than the minimum pulse width that N3 and N4 will reliably transmit. This imposes a restriction on the upper frequency of the circuit. As the frequency of the input signal increases, the duty-cycle of the output signal will approach 50% as the pulses come closer together. For this to happen, waveforms E and F have to have the minimum possible pulse width. In this situation the width of the positive output pulses is also the minimum which the 4069 hex inverter IC will handle.

Given the component values shown in the circuit diagram above, the width of pulses E and F is about 500 ns. Thus, when the frequency is 1 MHz, i.e., when the input frequency is 500 kHz, the duty-cycle of the output will be 50%.

14 THE HALF-WAVE RECTIFIER WITH A SHUNT CAPACITOR FILTER

A shunt capacitor filter circuit comprises a large valve capacitor which is connected in shunt with the load resistor. This kind of filter circuit is cheap and is the simplest that could be found.

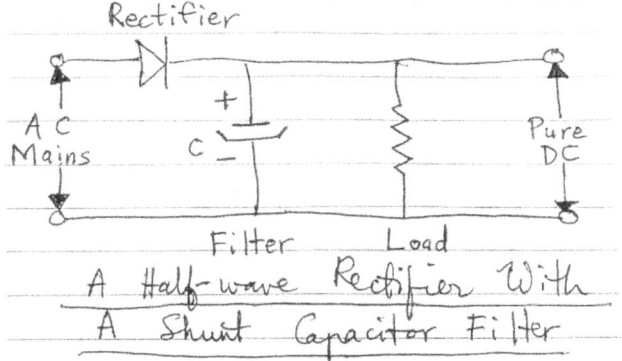

A Half-wave Rectifier With
A Shunt Capacitor Filter

The above diagram shows a half-wave rectifier with a shunt capacitor filter. The shunt capacitor presents a low reactance path to the AC components and the DC components. The diode conducts and the capacitor stores energy in the form of electrostatic field during the positive half cycle. The filter capacitor releases the stored energy to the load during the negative half cycle.

In the diagram below, the circuit's output waveform is shown:-

Output Waveform

15 HEAT SENSITIVE SWITCHES

In the circuit below, the thermistor's resistance is high when its temperature is low. As the potential difference between the base and the emitter is too low, the transistor is not switched on. The thermistor's resistance becomes low when its temperature rises. When this happens, the potential difference between the base and the emitter of the transistor becomes large enough to switch the transistor on. The lamp lights up.

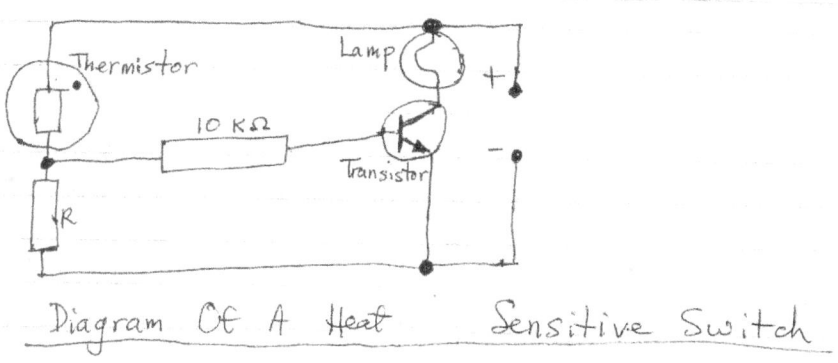

Diagram Of A Heat Sensitive Switch

16 INTEGRATED-CIRCUIT OSCILLATORS

Integrated circuits can also be used to construct oscillators. However, their performance is not as good as well-designed and well-constructed discrete-component oscillators.

An IC oscillator is shown below:-

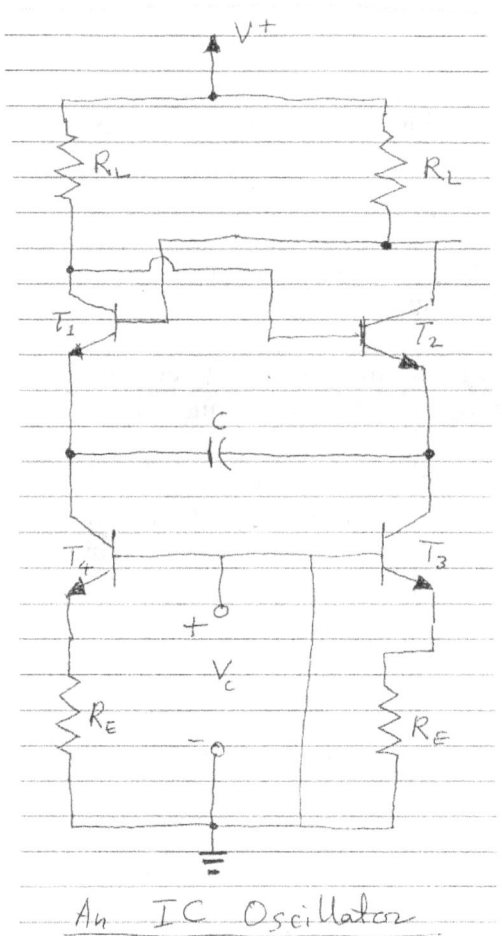

An IC Oscillator

The above IC oscillator does not use external components. It has a free-running multivibrator which uses transistors T_1 and T_2. Transistors T_3 and T_4 act as constant current sources. The current is proportional to the control voltage V_C as the emitter resistors R_E linearize the voltage-current relationship. Where the voltage V (approximately 0.6 V) is the voltage needed to turn the multivibrator off, the frequency of oscillation is:

$f = I_O/4CV$

The oscillating frequency is directly proportional to the control voltage as:

$I_O \approx V_C/R_E$

The output of the oscillator can be either a triangular or a square wave-form. However, to get a sinusoidal output wave-form, additional wave-shaping circuitry is needed.

It should be noted that as compared to IC oscillators discrete-component oscillators can be constructed with a better noise performance and can operate at higher frequencies.

A typical commercial crystal-controlled oscillator can operate between 0.1 and 20 MHz. The oscillator utilizes an external series crystal for precise frequency regulation. It has a wave-shaping circuitry which generates a sinusoidal output wave-form. There are also outputs available for directly interfacing with transistor-transistor-logic (TTL) and with emitter-coupled logic.

17 THE INTEGRATOR

A low pass filter is often called an integrator. This circuit removes the high-frequency components from a pulse wave-form, as the reactance of the capacitor falls with increasing frequency. The voltage across the capacitor cannot change instantaneously when a step input is applied. This voltage rises exponentially according to the following formula:-

$$V_C = V(1 - e^{-t/CR})$$

The time constant of the circuit is represented by CR, which is the product of capacitance in farads and resistance in ohms. The voltage across the capacitor changes by about 63% in one time constant. For the voltage across the capacitor to equal V, it takes nearly 4.5 time constants.

For pulses which have a long width in comparison to the integrator's time constant, the integrator would degrade the rise and fall times. The capacitor would not have sufficient time to charge completely and the output would appear triangular if the pulse is short in comparison to the integrator's time constant. To provide short time delays such circuits are often used.

Below is the schematic diagram of an integrator:-

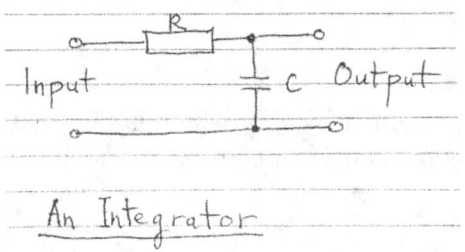

Input   C   Output

An Integrator

18 INVERTER CIRCUITS

The types of inverters, load, and, methods of voltage and current control determine the voltage and current ratings of power devices in inverter circuits. In designing inverter circuits, it is necessary to derive the expressions for the instantaneous load current and plot the current waveforms for each device and component. Establishing the reverse voltages of each device is necessary, when evaluating the voltage ratings.

Output filters are used to reduce the output harmonics. The commonly used output filters are shown below:-

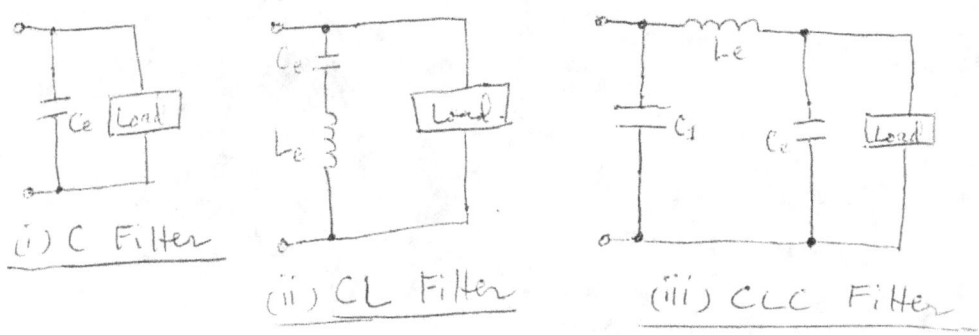

(i) C Filter

(ii) CL Filter

(iii) CLC Filter

The C filter draws more reactive power and is very simple. The LC-tuned filter eliminates only one frequency. The CLC filter draws less reactive power and is more effective in reducing harmonics of wide bandwidth.

19 INVERTER CONTROL CIRCUITS

Depending on the system used to vary the output voltage and the inverter configuration, a variety of control circuits exist for inverters. These control circuits comprise the power semiconductor drive circuit and a form of sequencing to turn the device on at the appropriate times in the cycle.

The following is a block diagram showing mark-to-space control within an inverter, using two shifted square waves:-

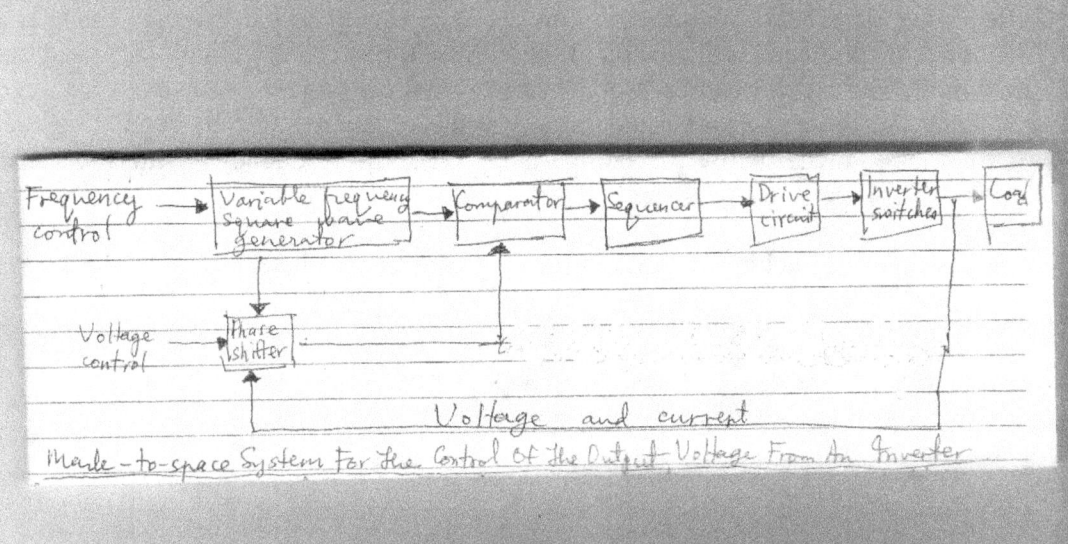

Mark-to-space System For the Control Of the Output Voltage From An Inverter

The square waveform from the generator is sent through a phase shifter. The comparator compares the two direct and shifted waveforms and provides the input to the sequencer and drive circuits. The drive circuit turns the inverter semiconductor switches on at the correct times. The load voltage is fed back and used to regulate the output. The current can be monitored to provide a current limit function, if required. By varying the frequency of the square wave generator and the output voltage by regulating the phase shift between the two square waves before the comparison stage, the frequency of the inverter can be controlled.

Selected harmonic reduction can be achieved by the following control circuit:-

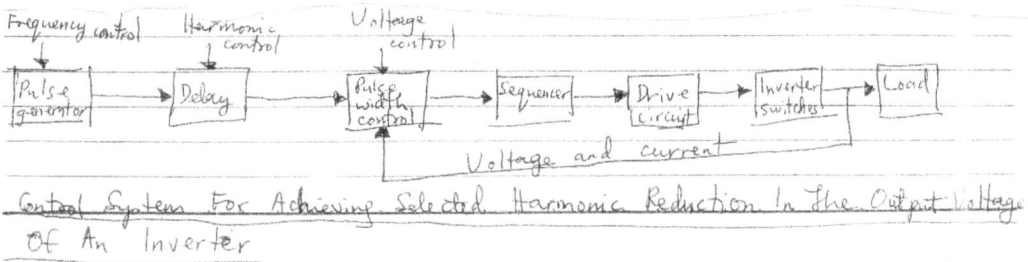

Control System For Achieving Selected Harmonic Reduction In The Output Voltage Of An Inverter

In this circuit the pulse generator output determines the frequency of the inverter. The pulse generator output is varied in order to produce a series of pulses which are spaced at the required distances apart to eliminate a given harmonic in the output voltage. By controlling the width of the pulses the magnitude of the fundamental voltage in the output is changed. As usual, the sequencer drives the correct power switches in the inverter while the drive circuit provides the drive power. The voltage and current feedback is used to sense and adjust the magnitude of the load voltage and current.

Below is a diagram showing a high-frequency pulse-width modulation control system:-

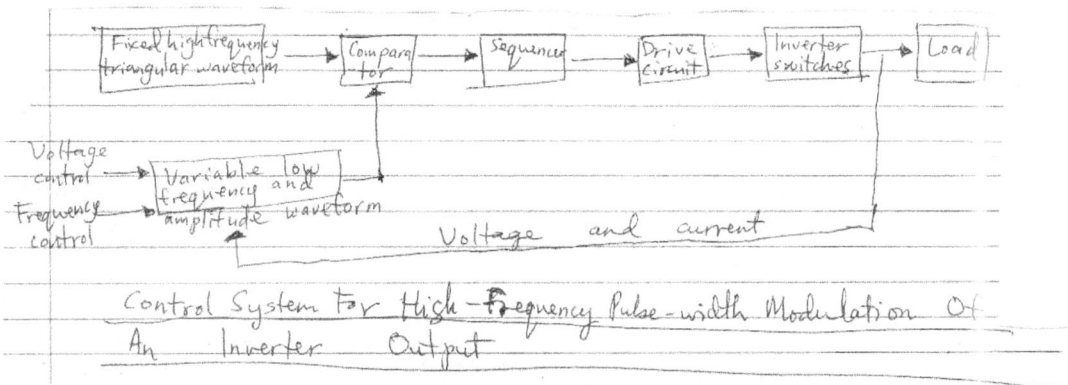

Control System For High-Frequency Pulse-width Modulation Of
An Inverter Output

The high-frequency triangular wave-form is compared with a low-frequency square, triangular or sine wave-form. The inverter frequency is determined by the frequency of this low-frequency wave. The magnitude of the output voltage is determined by its amplitude relative to the reference high-frequency wave. The rest of the circuit (sequencer, driver and voltage and current feedback) is as before.

The diagram below illustrates staggered phase carrier cancellation:-

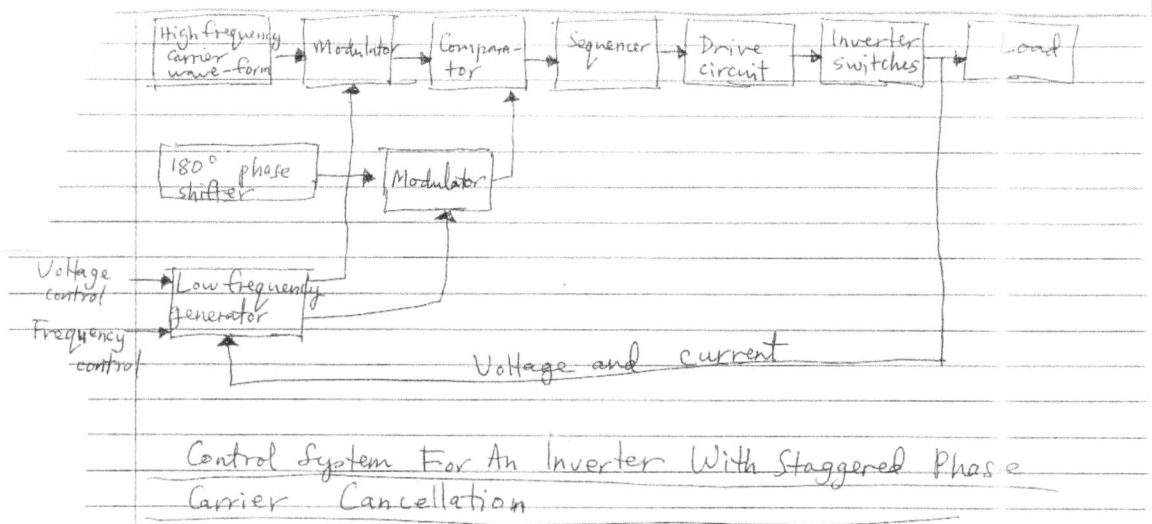

Control System For An Inverter With Staggered Phase Carrier Cancellation

Before being modulated by the low-frequency wave-form and then compared, the two high-frequency carrier wave-forms are phase shifted by 180°. All the subsequent stages in the circuit are the same as before.

20 INVERTING AMPLIFIERS

Negative feedback is utilized to bring the available gain of an operational amplifier (op amp) down to a useful value. A circuit which has this negative feedback is known as an inverting amplifier. The following is an inverting amplifier:-

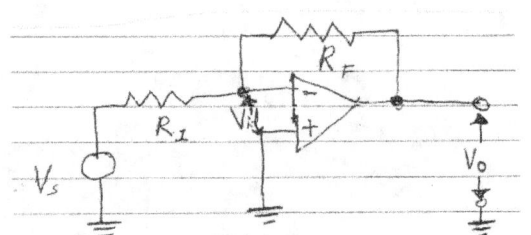

In the above circuit, resistors R_1 and R_F form a voltage divider. This causes a fraction of the output signal V_O to be dropped across R_1. The signal across R1, which is applied to the inverting input, is out of phase with the input signal V_S. It is called negative feedback. The stage gain, V_O/V_S, that results is smaller than V_O/V_{id}. In other words, if the stage gain $A_V = A_O/V_S$ and the open-loop gain $A_{VOL} = V_O/V_{id}$, then $A_V < A_{VOL}$. The stage gain A_V could be called the closed-loop gain.

The op amp such as that described above normally has a very large A_{VOL}. Compared to the input signal V_S, the signal V_{id} is very small. As V_{id} is very small, the inverting and non-inverting inputs are normally at the same potential. As the non-inverting input in the circuit shown above is grounded, the inverting input is also grounded.

The input resistances of op amps are very high. Because of this, most of the signal current flowing through R_1 is forced to flow through R_F. As the left side of R_F is virtually grounded the signal voltage across it is the output signal V_O. In other words, V_O is equivalent to $-R_F I$.

The output V_O is out of phase with the input V_S, as indicated by the negative sign in the above equation.

21 THE J-K FLIP-FLOP

Some feedback is involved from the output to the input when flip-flops are used as counting elements. As long as the clock input stays at logical 1, with edge-triggered systems there is bound to be oscillation between one state and the other. The master-slave system will help to improve this situation.

The diagram below shows how the J-K flip-flop combines the capabilities of the clocked S-R flip-flop with the master-slave principle:-

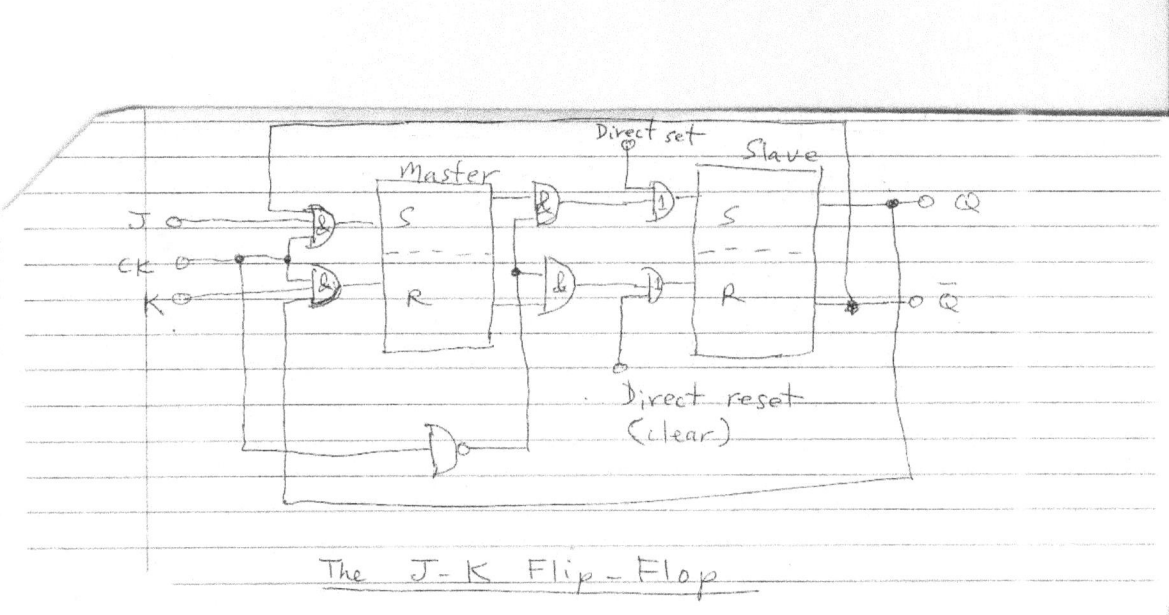

The J - K Flip - Flop

Shown below is the truth table for the J-K flip-flop:-

J	K	Q_{t-1}	Q_t
0	0	0	0
0	0	1	1
0	1	0	0
0	1	1	0
1	0	0	1
1	0	1	1
1	1	0	1
1	1	1	0

Truth Table For The J-K Flip-Flop

The circuit operates as a TRIGGER flip-flop when the signal applied to J = K = logical 1. In this situation, the flip-flop's output (Q) changes state at the end of each clock pulse as shown in the diagram below:-

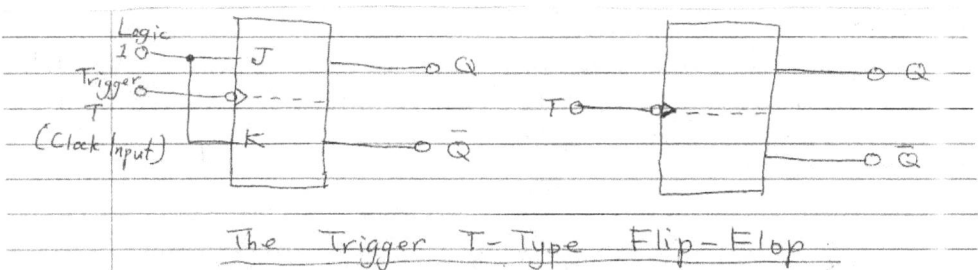

The Trigger T-Type Flip-Flop

As most of the desired facilities can be provided from clocked bistables, the J-K flip-flop is the most versatile. The D-type latch is readily available in various IC configurations and is the most common edge-triggered device.

22 LOOP FILTERS

Loop filters are low-pass filters which are usually of the first-order. However, when additional suppression of the AC components of the phase detector output is required, higher-order filters are utilized. Sometimes, a notch network is incorporated in the filter for suppression of a particular frequency. How the network could be configured depends on whether the phase-detector output could be modeled as a current source (high output impedance) or a voltage source (low output impedance). The diagram below shows a first-order filter that could be used with a charge-pump:-

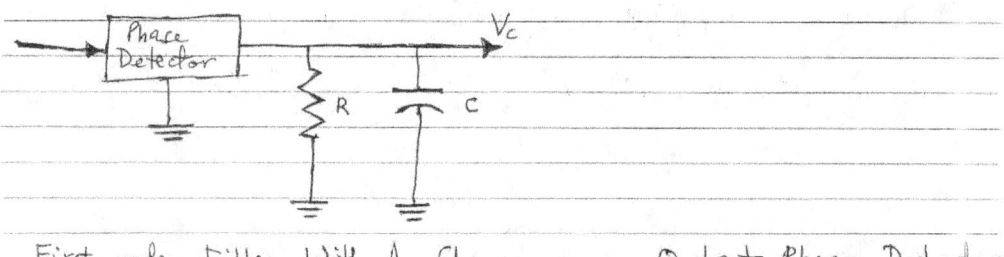

First-order Filter With A Charge-pump Output Phase Detector

The voltage is represented by:

$V_C = I(V)R/_sRC + 1$

$I(V)$ being the amplitude of the phase-detector output

A first-order active filter which could be used with a low output impedance phase detector is shown below:-

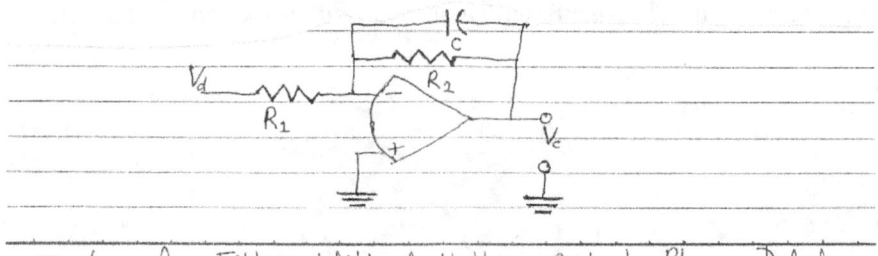

First - order Filter With A Voltage Output Phase Detector

The transfer function, which is a first-order filter, is as follows:-

$$V_C/V_d = -R_2/R_1[(sR_2C + 1)^{-1}]$$

Its DC gain, which could be adjusted to modify loop performance, is:

$$A_O = -R_2/R_1$$

When selecting loop filters, noise suppression, loop stability and transient performance should be taken into consideration.

23 MULTIPLEXERS

An electronic circuit that acts as a fast rotary switch is known as a multiplexer (time division type). The multiplexer connects several information channels, one at a time, to a common line. In a data acquisition system, the multiplexer is used to enable several sources of information to be sent along the common line and thus reduces the number of connections needed in any particular application. To produce analogue multiplexers, CMOS or JFET analogue switches or special-purpose ICs such as the LF 13508 could be used. Digital multiplexers, or, data selectors, could also be obtained in IC form. In digital multiplexers, the select channel address could be driven from a counter.

A demultiplexer at the receiver decodes data which is transmitted in multiplexed form. That is, the data is split out into the correct sequence of channel information.

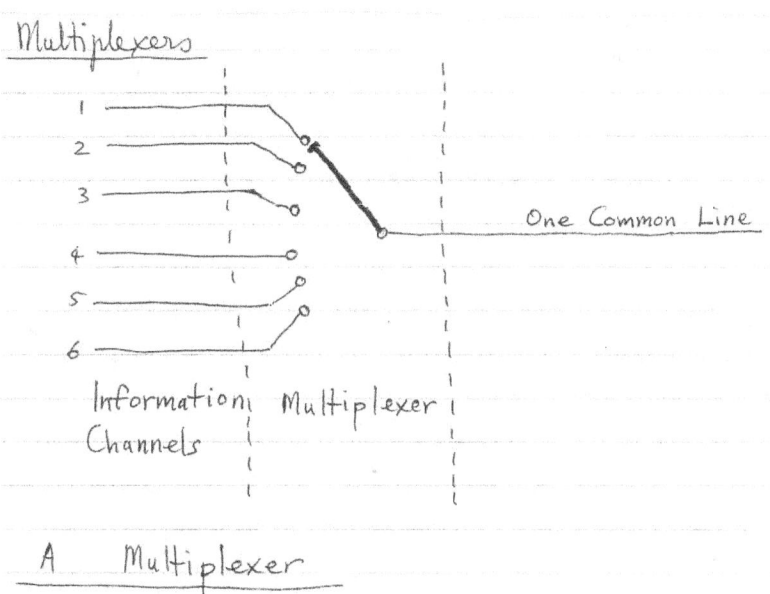

Multiplexers

Information Channels   Multiplexer   One Common Line

A   Multiplexer

24 OVERLOAD-PROTECTION CIRCUITS

Both current-limit and thermal-overload protection are evident in the diagram below. The current limit is set internally at approximately 2.2 A. It remains constant with temperature. When the chip temperature exceeds 170° C or so, thermal-overload protection, which is included in the chip, turns off the regulator. As the input-to-output voltage differential increases, the safe-area protection enables the pass transistor to decrease the current limit. Full output current could be achieved at 15 V differential, due to the safe-area protection. The safe-area protection does not allow the current limit to drop to zero at high input-to-output differential voltages. Start-up problems with high input voltages are thus avoided.

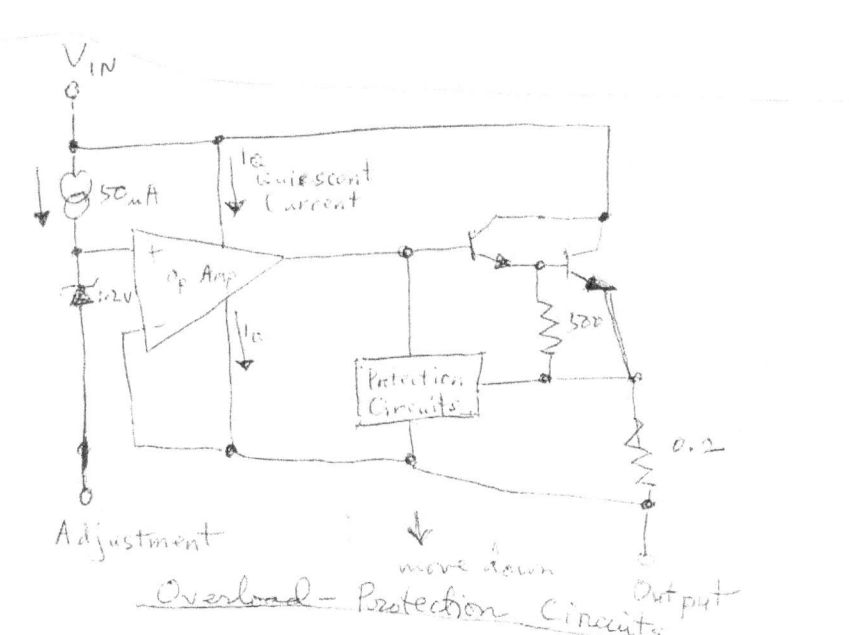

V_{IN}

$50\,mA$

Iq Quiescent Current

Op Amp

Vo

Protection Circuits

500

0.2

Adjustment

move down

Output

Overload - Protection Circuits

25 PHASE SHIFTERS

There are many methods of modulation and demodulation which require that the local oscillator signal be phase-shifted by 90°. Below is a circuit where a signal could be obtained together with the signal shifted in phase by 90°:-

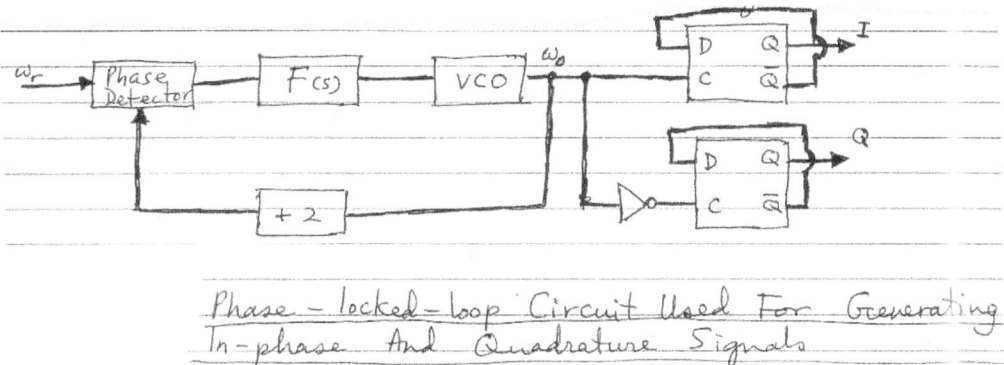

Phase — locked — loop Circuit Used For Generating
In-phase And Quadrature Signals

The reference frequency is doubled by the phase-locked loop. The flip-flop output frequency is similar to the reference frequency. As is evident from the timing diagram below, the two output signals differ in phase by 90° ($\pi/4 = 90^{\circ}$):-

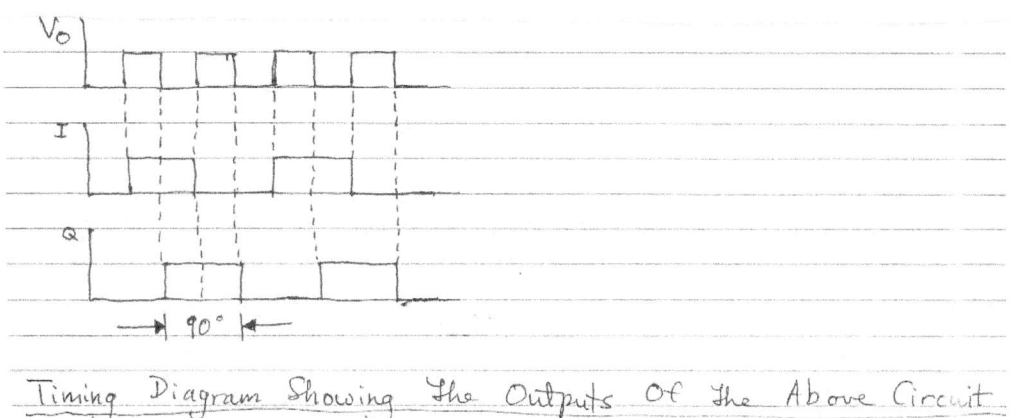

Timing Diagram Showing The Outputs Of The Above Circuit

26 PHASE-SHIFT OSCILLATORS

The phase-shift oscillator is an inductorless oscillator that is widely used. A typical phase-shift oscillator is shown below:-

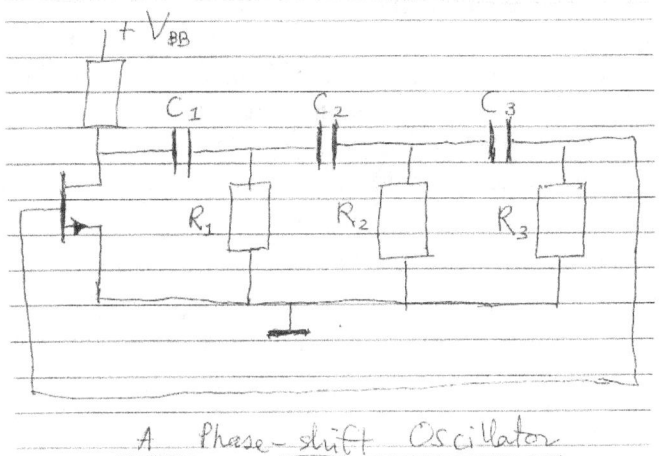

A Phase-shift Oscillator

The above resonator has three RC-networks. The RC-sections produce equal phase shift if $R_3 \gg R_2 \gg R_1$ and $R_1C_1 = R_2C_2 = R_3C_3 = T$. The total phase shift should be $180°$ if there were to be oscillation. This means that each RC-section has to contribute $60°$ phase shift. When oscillation occurs, w_o would be $1/(\tau\sqrt{3})$ while the attenuation would be $\frac{1}{2}$ per section. Therefore, $g_mR > 8$ must hold for the voltage amplification g_mR of the amplifier.

The biasing of the active devices was not taken into consideration in all the configurations mentioned above. The design of the biasing circuitry does not give serious problems generally. One terminal of the power supply is connected to ground for practical reasons. This also applies to the amplifying device. For oscillators that have to be tuned by hand one of the resonator terminals should also be grounded. A path for the biasing current should also be provided. We can attempt to configure the circuit in such a manner that the inductor is part of the dc path if the resonator has an inductor. If a capacitor has to be in series with a current-carrying electrode of an active device the capacitor could be bridged by a resistor. However, adding resistors to the resonator circuit would lower its Q.

27 PREAMPLIFIERS

There are applications in long-wavelength, long-haul routes for high input impedance preamplifiers. This type of preamplifier is a highly sensitive one, due to the use of a high input resistance preamplifier (typically > 1MΩ) which gives rise to exceptionally low thermal noise. Combined with the receiver input capacitance, the high resistance gives rise to a very low bandwidth, which is typically < 30 kHz, and this results in integration of the received signal. These receivers are commonly called "integrating front-end" designs. This integration is corrected by a differentiating, equalizing or compensating, network at the receiver output.

Either field effect transistors (FETs) or bipolar junction transistors (BJTs) could be used by both types of preamplifier as the input device. Usually, FET input receivers are more sensitive than BJT input receivers. But, the situation can change at high data-rates (typically greater than 500 MHz). For the integrating front-end receiver we shall look at a FET input design, while for the transimpedance receiver we shall use a BJT.

Preamplifiers With High Input Impedance
The choice of a front-end transistor is important since these preamplifiers rely on a very high input resistance to produce a sensitive receiver. BJTs are seldom used because they have a relatively low input resistance. FETs are the obvious choice for the front-end device because they have a very large input resistance. Integrating front-end preamplifiers usually have a PIN photodiode feeding a FET input preamplifier. The resulting circuit is known as a PINFET receiver. The basic PINFET receiver design is shown below in simplified form:-

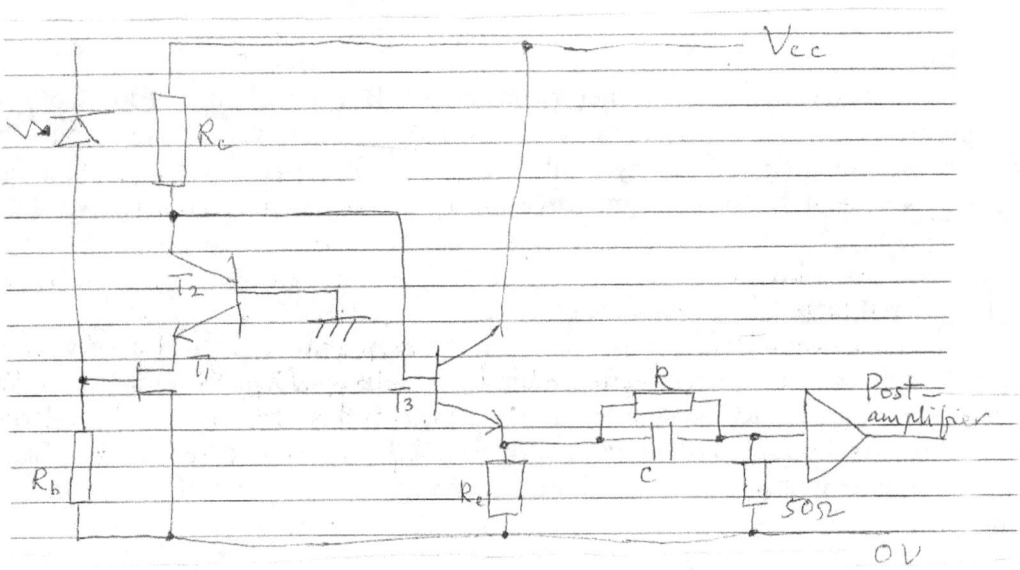

Basic PINFET Optical Receiver With Equalisation
Network And Post-amplifier.

28 PROPORTIONAL-INTEGRAL-DERIVATIVE (PID) CONTROLLERS

As is implied by the name, proportional-integral-derivative (PID) controllers utilize the attractive attributes of all the three controllers, viz., the proportional controller, the integral controller and the derivative controller. The proportional aspect gives fast response to system disturbances, the derivative part makes sure that sudden disturbances would be met with an aggressive effort to correct the error, and the integral portion gives a means of finally getting rid of the error altogether.

The diagram below shows a common parallel configuration, although there are many possible PID variations:-

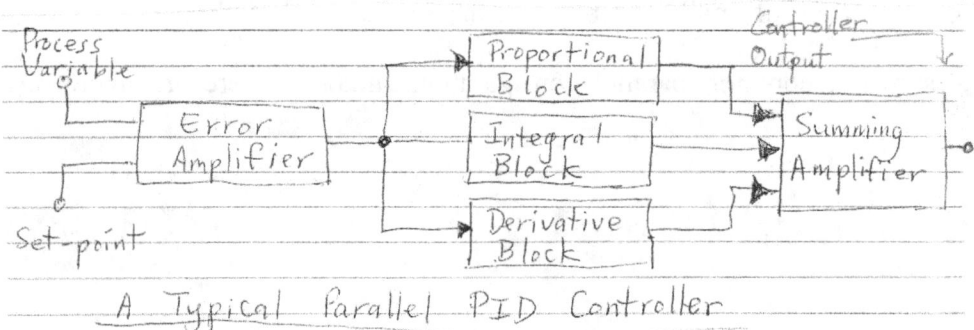

A Typical Parallel PID Controller

The same error signal is received by each element. A summing amplifier adds the outputs of all the elements. So long as the error is represented by a relatively simple function, it should not be difficult to predict the output response to a change in error signal.

Tuning is the process of adjusting each of the three blocks in a PID controller. The tuning of a PID controller depends on the characteristics of the process being controlled, the desired controller performance and the controller's configuration. Each PID controller would require a different tuning procedure though the same controller configuration is applied to two different processes. The tuning process is not an easy task. As a guideline, literature published by the controller manufacturer is frequently used. However, computer simulation programs are popular as results could be observed quickly without having to start up the process though the accuracy of the results depends on how well the response of the system could be modeled.

When executing PID control two precautions should be observed. This is because the activity of the integral or derivative block can mask the effects of the other blocks in the controller. For instance, the derivative block would most likely saturate, causing a corresponding saturation in the summing amplifier, if there were a sudden change (step) in error (which could be caused by a change in the set-point or a disturbance in the process). As a result, there might be an overcompensation which causes oscillations in the process. The integral block output might be forced into saturation if a huge error were present in the system for a long period of time. The integral block output would remain at saturation even if the error were driven back to zero, and would cause the process to overshoot until the resultant negative error brings the integral block out of saturation.

29 THE Q METER

The diagram below depicts the circuit of the Q meter. In this circuit, a RF oscillator which generally covers the frequency range from 50 KHz to 70 MHz in different ranges is used. A small voltage which is precisely calibrated is supplied from this oscillator to the circuit which comprises the coil under test. An electronic voltmeter that is of high input impedance and calibrated directly in terms of values of Q indicates the voltage across the internal capacitors.

Up to 20 MHz, quite good accuracy of the order of ± 2.5% to ± 5% could be achieved.

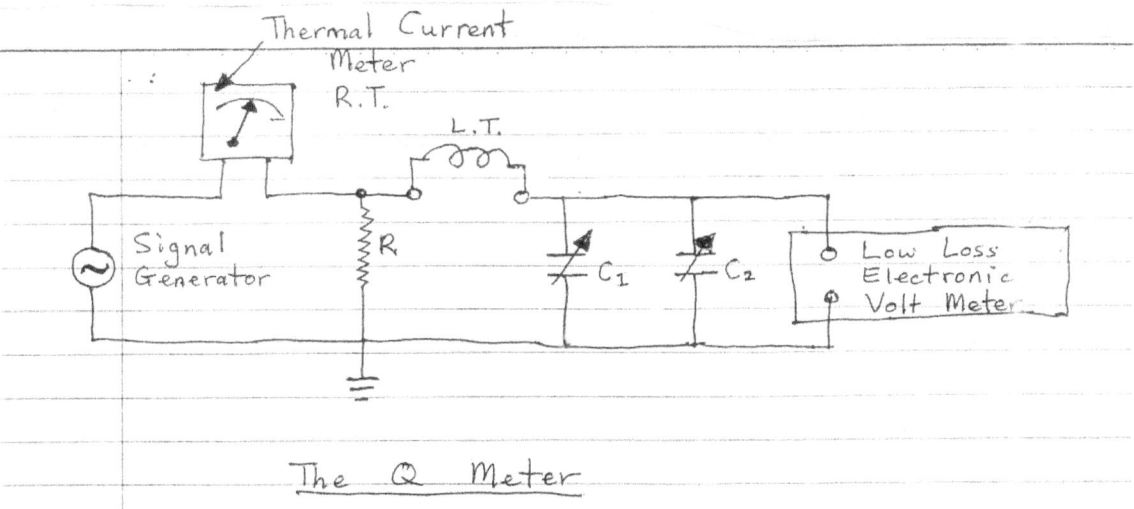

The Q Meter

30 THE RAMP ANALOG TO DIGITAL CONVERTER

The simplest type of analog to digital converter (ADC) is the "ramp" kind. The number-selection method is very simple. A counter counts from zero upwards. A comparator finally informs the logic that the unknown and the digital to analog converter (DAC) output have passed each other. The ramp ADC circuit is shown below.

The advantages and disadvantages of the ramp ADC are as follows:-

[a] With a DAC, it is very easy to implement.

[b] However, as the DAC has to slew from zero to the unknown voltage value and the counter has to count from zero to the answer, it can be slow.

[c] As the number (and its analog) represent a value that is greater than the unknown by up to one LSB, instead of being ± ½ LSB on the unknown, it is important that a ½ LSB offset be introduced before the comparator.

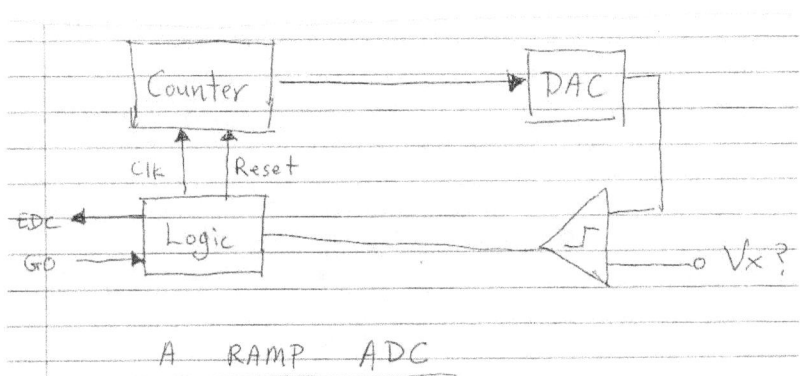

A RAMP ADC

31 THE SET-RESET FLIP-FLOP SWITCH DEBOUNCER

Memory circuits in general and counting circuits in particular face the problem of differentiating between real signals and glitches. When the contacts are closed, mechanical switches suffer from bounce, resulting in a circuit that is attached to a switch seeing an input wave-form like that for Š or Ř, as shown in the diagram below, where the oscillations could last several milliseconds:-

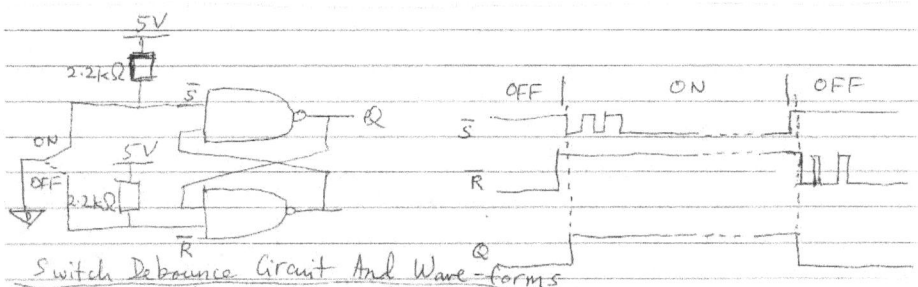

Switch Debounce Circuit And Wave-forms

A circuit that recognizes the transition but ignores the bounces is needed. This could be provided by an SR flip-flop and an SPDT (single-pole double-throw) switch. The switch should open one set of contacts prior to closing the other, so that $\check{S}$ and $\check{R}$ would not be zero simultaneously. A bounce should not occur on the contact that is opened, as that would appear as a real switch operation. These requirements could be met by most simple toggle switches. In order to minimize the current flowing through the switch, NAND gates and pull-up resistors could be incorporated in the circuit. A resistance of a few hundred ohms is needed to establish a logical zero at a TTL gate input for a circuit with NOR gates and pull-down resistors.

32 STEPPER MOTORS

Stepper motors are digital devices which convert electric pulses into proportionate mechanical movement. Each revolution of the stepper motor's shaft consists of a series of discrete individual steps. The stepper motor usually caters for both clockwise and counterclockwise rotation. In the industrial world, it has a wide variety of control and positioning applications, e.g., in robotics, peripherals, instrumentation controls and machine tools. Its usage has increased rapidly and continues to do so with the fast growth of solid-state electronics and digital techniques.

The rotation of the stepper motor shaft is incremental. On being energized it moves and comes to rest after some number of steps in strict accordance with the digital input commands provided. It allows control of the load's velocity, distance and direction and is able to position through the same pattern of movements a number of times (repeatability).

The stepper motor is a basically reliable device. The only part of the motor subject to wear is the bearings. It is a good substitute for many shorter-lived, more maintenance-intensive devices such as clutches, gears and brakes, as it gives an overall improvement in reliability.

There are three main types or classes of stepper motor, each having its own distinct construction and performance characteristics, viz., variable-reluctance (VR), permanent magnet (PM) and PM-hybrid.

The operation of a stepper motor depends on the basic magnetic principle - similar magnetic poles repel one another and dissimilar poles attract one another. A stepper motor is shown below:-

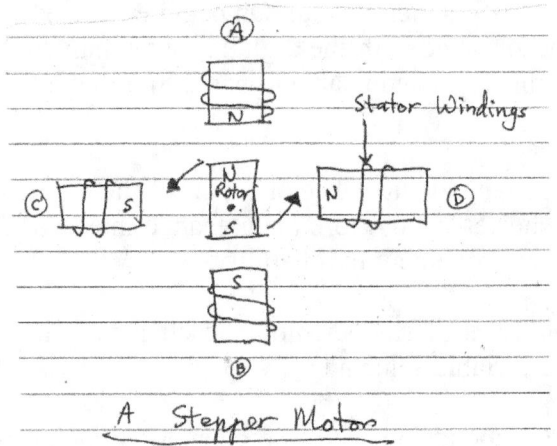

A Stepper Motor

In this stepper motor, the stator windings are energized such that stator A is the north pole, stator B is the south pole, and the permanent magnet (PM) rotor is positioned as indicated. As shown, two more stator poles C and D are added and energized. The rotor would rotate in the counterclockwise direction.

As stepper motors could be used in an open loop mode while still offering many of the desirable features of the feedback-type system they are popular. They are excellent positioning devices, offering great reliability and consistency, reasonable cost and consistent performance. However, there are some limitations. They are not very energy-efficient. The available torque is inversely proportional to the speed. If the stepper motor were commanded to go from stop to full speed immediately it would stall - speed has to increase gradually. They have a low speed resonance point, whereby the torque is reduced drastically.

33 SWITCHED-MODE POWER REGULATORS

As compared with generally less than 50% for conventional regulators, switched-mode regulators can operate with efficiencies of 80% to 90%. As heat dissipation is at the minimum, switched-mode regulators can be made extremely compact. However, the circuitry used is rather more complicated than that linked with more conventional linear regulators.

The diagram below shows a switched-mode regulator based on discrete circuitry, that provides 5 V at up to 1 A:-

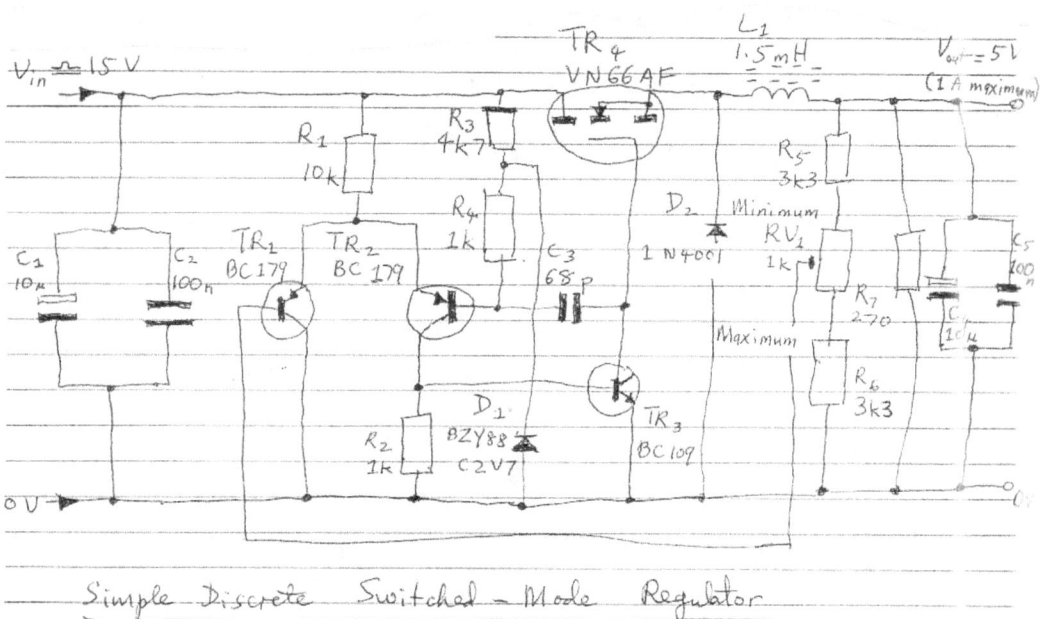

Simple Discrete Switched – Mode Regulator

The above circuit operates at a switching frequency of about 125 kHz. It has an output resistance of less than 0.2 Ω. Its output voltage adjustment range is from 4.6 V to 5.8 V (approximately). The FET, TR_4, is a V-MOS power type transistor which has a maximum rated drain current of 2 A. The inductor, L_1, is wound on a high-permeability ferrite pot core.

The dc output of the circuit can have a significant proportion of high frequency noise. Therefore, cautious attention should be given to adequate decoupling, and, where necessary, screening of the entire power supply module. At least two capacitors of widely differing values and appropriate construction (e.g., 10 μF radial lead PCB electrolytic and 10 nF disc ceramic) should be used for decoupling. The input to the regulator should be like-wise decoupled to prevent radiation of noise from the rectifier and transformer wiring, as the main reservoir capacitor is ineffective for decoupling at the relatively high switching frequencies concerned.

34 VOLTAGE MULTIPLYING RECTIFIERS

In order to produce higher output voltages, various circuits in which capacitors are connected in series and charged via diode rectifiers can be used.

A circuit which acts as a voltage doubling rectifier is shown below:-

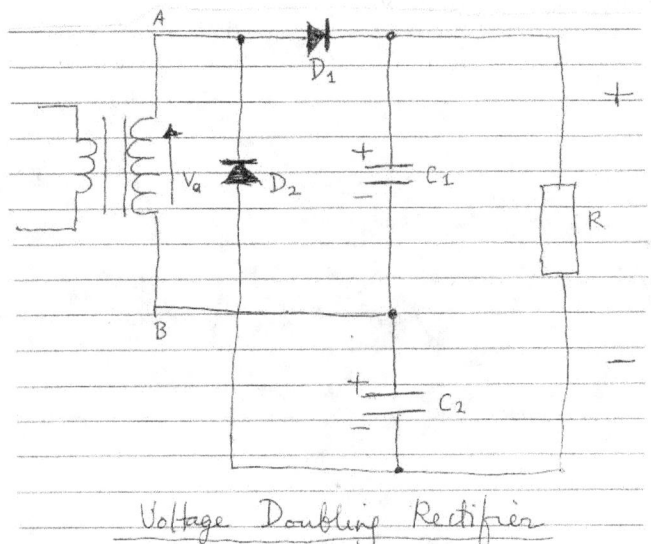

Voltage Doubling Rectifier

Capacitor C_1 charges when A becomes positive with respect to B, with the polarity shown via diode D_1. In the next half cycle, capacitor C_2 charges with the polarity shown via diode D_2 when B becomes positive with respect to A. The load voltage is taken from across the two capacitors connected in series and is the sum of the two voltages. Each capacitor would charge to the peak value of the voltage V_a if the load R is disconnected - the direct output voltage would therefore be equal to $2 V_a$.

35 VOLTAGE REGULATORS

By using sufficiently large capacitors, the ripple amplitude of a dc current, obtained by rectification of an ac, could be reduced to any desired level. But, this approach presents several disadvantages which are as follows:-

[1] There might be a need for exceptionally bulky and expensive capacitors.
[2] The output voltage changes with the load, due to the finite internal resistances of the elements of the rectifying and filtering circuits (transformer, resistor, diode, et al.).
[3] Input line fluctuations are transmitted to the output voltage.

To solve this problem, a regulator could be used together with a capacitor.

Regulators give a constant dc voltage and are active feedback circuits. Voltage regulators are very commonly used as power supplies in electronic circuits. They are available as inexpensive chips. As a voltage source, a power supply built up with a voltage regulator has excellent properties. To get the desired voltage, voltage regulators could easily be adjusted. They are internally self-protected against short circuits and overheating.

A voltage regulator is shown below:-

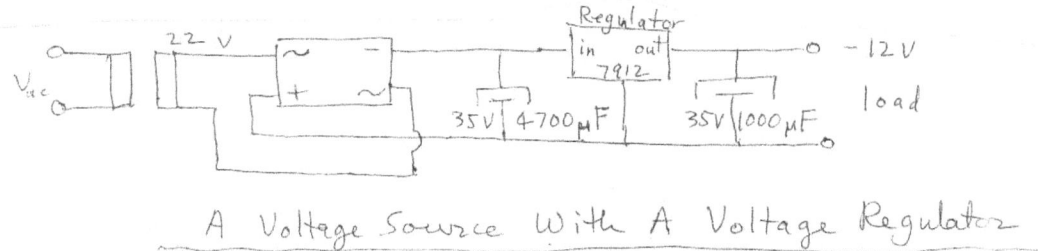

A Voltage Source With A Voltage Regulator

At the cost of some power, which is dissipated as heat by the voltage regulator through the input-ground path, the load, connected to the output and ground terminals of the regulator, is kept at a constant voltage. The capacitor across the output further diminishes the residual ripple.

36 VOLTAGE-TO-CURRENT CONVERTERS

The converter shown below, which is also known as voltage-controlled current source, has only one operational amplifier and draws from and provides to ground a current that is dependent on its input voltage:-

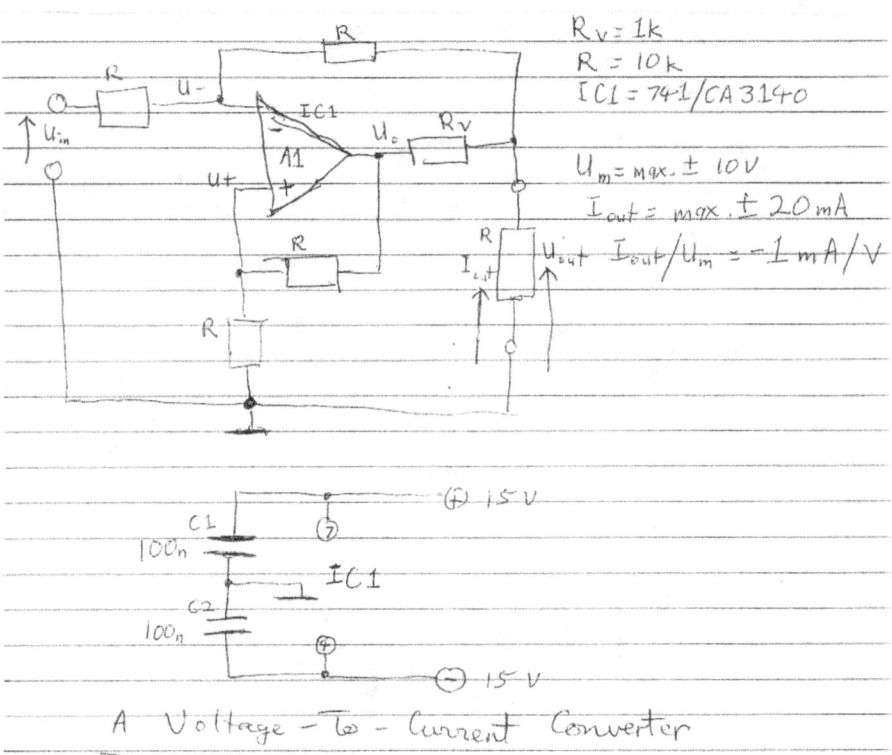

$R_V = 1k$

$R = 10k$

$IC1 = 741/CA\,3140$

$U_m = max. \pm 10\,V$

$I_{out} = max. \pm 20\,mA$

$U_{out} \quad I_{out}/U_m = -1\,mA/V$

A Voltage - To - Current Converter

The above circuit could convert both negative and positive voltages into negative currents (from ground) and positive currents (into ground) respectively.

The operational amplifier (A1) could be a Type 741 or CA 3140, whereby $R_v = 1k$, or $R = 10 k$, $U_{in} = \pm 10$ V max., $I_{out} = \pm 20$ mA max. and gm = -1 mS. Some or all of these values could be varied as required by using a different operational amplifier and changing the values of the resistors. The maximum output current always depends on the operational amplifier used. The following formulae could be useful for such changes:-

$U + = U - = (U_{in} - U_{out})/2 + U_{out}$

$U_o = 2[(U_{in} - U_{out})/2 + U_{out}] = U_{in} + U_{out}$

$I_{RV} = U_{in}/R_v$

$I_{out} = I_{Rv} + I_R = U_{in}/R_v + (U_{in} - U_{out})/2R$

If $R >> R_v$ (which is the usual case),

$I_{out} = U_{in}/R_v$

37 WAVE-SHAPING CIRCUITS

Operational amplifiers can provide output voltage that is proportional to the differential function or integral function of the input voltage. These two circuit configurations can be regarded as a means of performing simple wave-shaping operations.

Below is the circuit of a typical operational differentiator:-

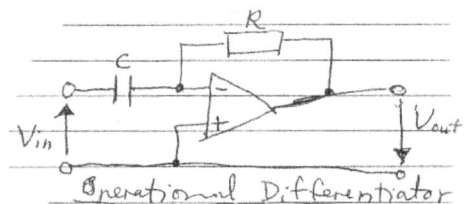

Operational Differentiator

The circuit below is that of a typical operational integrator:-

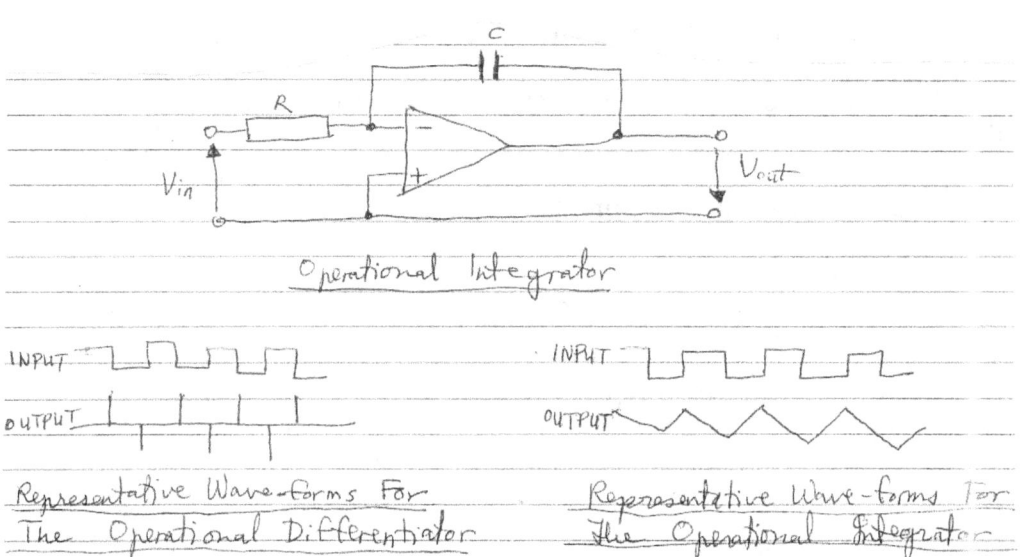

Operational Integrator

INPUT

OUTPUT

INPUT

OUTPUT

Representative Wave-forms For
The Operational Differentiator

Representative Wave-forms For
The Operational Integrator

Representative input and output wave-forms for the operational differentiator and the operational integrator are shown above.

To achieve effective differentiation of a wave-form by the operational differentiator, the following condition has to be met:-

t << C x R (where t is the periodic time of the input voltage)

For the case of an operational integrator, effective integration of a wave-form can be achieved by meeting the following condition:-

t >> C x R (where t is the input voltage's periodic time)

38 GUIDE TO ELECTRONIC CIRCUIT DESIGN PART I

1.1   The electronics engineer who designs electronic circuits relies very greatly on data-books from IC and component manufacturers.

1.2   These data-books are normally obtained free-of-charge.

1.3   The designer normally keeps a look-out for the latest ICs available in the market.

1.4   Vendors who supply the ICs would usually keep him posted on the latest available products and technology.

1.5   The rationale is that the company has to keep up with the competition.

1.6   The electronics engineer therefore has to know what the latest technology in the market is - they have to be better than the competition, or, at least, be on the par with them.

1.7   Faster ICs, with more memory capacity, and other state-of-the-art components/parts such as self resetting power supply cut-off switches (in lieu of fuses) and "8-layer" PCBs, are a boon to the electronics engineer.

1.8   IC data-books show ICs with their related electronic circuits and they provide a host of technical information such as current requirement and input and output voltages.

1.9   Thus, nowadays, the electronics engineer seldom has to design a circuit from scratch as such circuits are freely available for adoption in the data-books.

2.1   However, not all electronic circuits shown in IC data-books would work excellently according to specifications.

2.2   The engineer normally has to make a proto-type of the circuit he has selected and test it out.

2.3   He might have to modify the circuit a little, e.g., by varying the values of a component or two here and there.

2.4   The electronic circuit could only be considered successful, when, on testing, is found to work according to specifications.

2.5   However, a circuit might fail, not because of an inherently faulty design, but due to careless workmanship, such as poor soldering.

3.1   Nowadays, a whole electronic circuit on a PCB could be "compressed" into an IC.

3.2   It pays to look out for ICs which could replace a whole PCB with its many assembled electronic components such as resistors, diodes and capacitors.

3.3   This would help cut down design-time and make the final product more efficient.

3.4   There are organizations which set out to custom-design such ICs.

4.1   Other technologies, e.g., surface-mount-technology, could make the final product smaller, more compact and more efficient.

4.2   Examples of products utilizing some of these latest technologies include lap-top and palm-top computers and laser disc players.

4.2.1   For instance, computers nowadays are not only smaller and take up much less desk space, they are faster in operation and have more interesting features and memory capacities.

5.1   The more ambitious electronic circuit designer might want to design a new circuit from scratch.

5.2   He might have a brilliant concept or two and might spend many long hours and week-ends trying to perfect his very own design (by testing and re-testing his proto-type and checking and rechecking his calculations).

5.3   If his design is found to work out well, he could end up with a patent for it.

6.1   In short, the electronics engineer should always be alert and on the look-out for the newest devices and components/parts that are available in the market.

6.2   He should try to cultivate a good working relationship with the vendors of devices and components.

6.2.1   He should be able to tap their knowledge and expertise.

6.3   He should be curious and keen to learn new things.

6.4   He should maintain a good library of component data-books and other reference books.

6.5   The component data-books would offer a wide range of electronic circuits, e.g., audio, video and transistor circuits, for him to choose.

6.5.1   For instance, different semi-conductor manufacturers offer various devices with differing electronic characteristics and values and the indicated electronic circuits incorporating these devices also vary.

6.6   The electronics engineer has therefore to use his discretion, based on his experience and knowledge and on the cost/benefit factor, to select the device and electronic circuit that would deem to best suit his requirement.

39 GUIDE TO ELECTRONIC CIRCUIT DESIGN PART II

1.1   Proficiency in circuit design would come with experience.

1.2   The author would like to provide some pointers to the budding circuit designer, which are as follows:-

[1]   Keep a proper documentation (make proper notes) of every stage of the design. Any changes in the design must be documented. This facilitates fault-tracing when things do not turn out right.

[2]   Keep a note-book (for reference) of useful circuits (which you have designed and used effectively in the past). Always refer back to this note-book when working on the design.

[3]   Have a well-stocked library of integrated circuits (ICs) data-books from the various semiconductor manufacturers, e.g., Texas Instrument, NS, Matsushita, et. al., as well as data-books on other electronic parts such as diodes, capacitors, transformers, et al. You might have to be selective of the parts to be used and such data-books could provide a wealth of useful technical information. For example, a whole lot of electronic parts in a certain stage of your design might be substituted by a simple IC, after carefully studying the data of the selected IC in the data-book.

[4]   Always keep an open mind. No two circuit designers would come up with the same design.

[5]   Try to minimize on utilization of components, parts for cost-saving, time-saving and design efficiency purposes.

[6]   Maintain a library of reference books on electronic circuits, for easy and quick reference, as well as electronics journals with electronic circuit ideas.

[7]   Intermingling and discussions with fellow professionals, during professional society activities and/or seminars/conferences, would help boost circuit design creativity.

[8]   Always try to keep up-to-date by reading widely, attending courses and talks, visiting trade shows and exhibitions, et al.

[9]   Remember that in a dynamic specialization such as electronics it is necessary to be prepared for life-long learning. (Thus, you should have developed your learning skills.)

[10] Nowadays, advanced software such as Orcad and Mentor Graphics allow you to design, simulate and test the circuit (you have designed) by the computer alone (which means that you do not have to construct a physical model based on your design to test (by using the multimeter, oscilloscope, logic probe and what-not)). Therefore, it is important to keep up-

to-date with computer technology and software.

1.3   The author would like to assure the reader that with experience, enthusiasm for learning and exposure a good circuit designer is likely to result.

40 EPILOGUE

1.1 Modern electronics can be split into two main branches: analog electronics and digital electronics.

1.2 The whole field of digital electronics depends on the principle of transistor switching.

1.3 Digital circuits are just several simple transistor switching circuits that are used many times over, e.g., those found in a computer.

1.4 The basis of all analog or linear circuits is the transistor amplifier. (Most analog circuits are made up of several transistor amplifiers, and are currently based on the IC op-amp (operational amplifier).

1.5 The other aspect of electronics that has been touched on here is related to the non-amplifying electronic components which are called passive components.

BIBLIOGRAPHY

1) Horowitz and Hill. The Art Of Electronics. Cambridge University Press. 1980.
2) Jones. A Practical Introduction To Electronic Circuits. Cambridge University Press. 1981.
3) Melville. Electricity. The Hamlyn Publishing Group Limited. 1978.
4) Worcester. Electronics. The Hamlyn Publishing Group Limited. 1980.
5) Duncan. Adventures With Electronics. John Murray. 1978.
6) Duncan. Adventures With Microelectronics. John Murray. 1981.
7) Duncan. Adventures With Digital Electronics. John Murray. 1982.
8) Sinclair. Practical Electronics Handbook. Newnes Technical. 1980.
9) Pierce and Paulus. Applied Electronics. Merrill. 1972.

www.ingramcontent.com/pod-product-compliance
Lightning Source LLC
Chambersburg PA
CBHW081726170526
45167CB00009B/3711